WHAT THE F*CK IS CYBER?

First edition - First published by Tidy Cyber, LLC 2020

Copyright © 2020 by Jae Tyler

Editing by Rosie Manins

Editing by Quentin Fields

Cover art by Tyler Hann

Advising by Kevin Grimes

All rights reserved. No part of this publication may be reproduced, stored, or transmitted in any form or by any means, electronic, mechanical, photocopying, recording, scanning, or otherwise without written permission from the publisher. It is illegal to copy this book, post it to a website, or distribute it by any other means without permission.

Jae Tyler asserts the moral right to be identified as the author of this work.

Jae Tyler has no responsibility for the persistence or accuracy of URLs for external or third-party Internet Websites referred to in this publication and does not guarantee that any content on such Websites is, or will remain, accurate or appropriate.

Designations used by companies to distinguish their products are often claimed as trademarks. All brand names and product names used in this book and on its cover are trade names, service marks, trademarks, and registered trademarks of their respective owners. The publishers and the book are not associated with any product or vendor mentioned in this book. None of the companies referenced within the book have endorsed the book.

Paperback - ISBN:	978-1-7357860-0-1
eBook - ISBN:	978-1-7357860-2-5
PDF - ISBN:	978-1-7357860-1-8

for my

Grandmother Mamie

Grandmother Faye

Grandmother Pearl

Grandmother Roxie

FOREWORD by Quentin Fields

What the F*ck is Cyber is a modern, how-to guide that aims to assist individuals with navigating the complexities of everything related to cyber. The author implements an informal approach to writing that feels similar to an everyday conversation; instead of traditional instruction taught in grade school and higher education institutions. This approach fosters a mindset that establishes expectations and encourages the reader to actively engage the contents of the book. ***What the F*ck is Cyber*** has something to offer to a range of audience members with varying degrees of experience. For the novice, the book provides a solid foundation on the basic components of cyber. For the intermediate reader, the book serves as a refresher on traditional concepts such as the decimal system, networking protocols, and programming languages. For the expert reader, the author provides useful ways to efficiently incorporate technologies to improve professional digital workspaces and personal lives (refer to the ***DIGITAL MINIMALISM*** chapter). By the end of this book, every reader, at all levels, should be able to answer the question, ***What the F*ck is Cyber.***

PREFACE

This book was crafted for two distinct purposes. One was to keep a promise to my grandmother who has passed. I told her when I was much younger. I would write a book some day and she always asked about it. The second was to create a book for people who are not fully computer literate, like my still living grandmother. This was made for them.

I have worked in many industries at varying levels. I have been a cyber security instructor, software engineer, technical consultant, website designer, and even a director of social and digital media. Though, my college education is brimming with social sciences; as my formal degrees are in psychology and international studies. I am a bit of an odd duck.

Through my experience in both the social sciences and my depth of experiences in technology I realized something. Life surrounding technology should be demystified a bit. Thus, the concept was birthed of *What the F*ck is Cyber to the average user*.

Please give this book a fighting chance. It is my first and like most first. There may be some rough patches. I, also, want to say thank you for your kindness and understanding in advance.

INTRO

DON'T STRESS

If you are ever having self-doubt or concern about cyber, computers, or anything information technology (IT) related, then just look at the back cover of this book. Keep in mind, this book is meant to be a light read. This is merely a guide on a continuous journey of computer literacy.

WHAT THE F*CK IS CYBER

Cyber is everything related to computers, technology, digital devices, and the culture binding them together. This age of cyberspace has a lot of depth. It is ever evolving and always progressing. To what? I personally do not know. But I know this: it is vital to learn to ride the wave and not get lost in it.

IS THIS BOOK FOR ME

Have you ever felt the need to take a break from social media? Ever wondered why you keep getting a 'storage is full' message on your phone? Is your desktop littered with icons and files? Or is your phone cluttered with apps you never use? Do you cycle through the same three or four apps kind of like you are opening a fridge when you are hungry only to find the contents have not changed? Ever feel overwhelmed with technological advances? Or are you scared technology is moving faster than you can keep up with? Then you should feel right at home. You are in the right place. This book is for you, I promise.

CALL ME YOUR SHERPA

Computer technologies can be confusing, and I used to actively avoid learning about them. I was lost and clueless about anything related to computers for most of my life, until a few random career changes after college led me into the technology field.

Forced to learn and become endlessly curious about computer science and all things related, I'm now an average guy with a ton of cyber experience, who wants other tech beginners to have an easier path to getting used to digital life. So, call me your cyber Sherpa.

WHAT'S IN THIS BOOK

I am not going to bore you with too much history or dive deeper than what is necessary to get a point across. There are tons of resources already taking a microscope to all thing's computer technologies. We are going to take a step back and look at the whole picture.

Many books already focus on the technically minded -- super geeks if you will -- and computer engineers. I feel few books focus on the everyday person. I am here to equip you with simple concepts to understand the complex universe of cyber.

My ultimate goal is to help beginners become computer literate. Essentially, I want even my grandparents thinking and talking cyber.

THE FLOW OF OUR JOURNEY

Our journey together starts with understanding the mindset required to learn and grow cyber knowledge. Cyber has a massive range of topics to cover from agriculture to education to global monetary systems. I cannot cover the full scope within the contents of this book, no book can. However, I have hope and believe, that discussing the mindset required will provide you with the ability to continue your cyber journey beyond the content I present in this book.

After addressing the mindset it takes to better understand computer technologies and our changing world, I will be touching on a variety of topics; such as the basics of counting and how it relates to computers, the internet and its impact on our lives, the inner workings of a computer and what a computer is, and cyber security. These topics are vital to developing a strong foundation in the world of cyberspace. They are the building blocks understood by every software engineer, network administrator, and cyber security professional. They are equally important to every profession, persons, and walk of life. Understanding the core of technology will help you to be more efficient, adventurous, curious, and comfortable in the digital world we live in.

I will also address more holistic topics like: digital minimalism, decluttering your digital space, creating technology free spaces, and ways to use technology without being controlled by it. I am discussing these topics because computer technology is more than 1s and 0s. The effect technology has on our lives can result in a positive or negative influence of our daily experiences. I think it's best we explore the dynamics of technology while developing the tools and skill sets to address its ever-growing presence.

The last section of this book is a bit free form. The topics covered are all related to information from previous chapters, but I could not find a perfect place for them. There will be an exploration of artificial intelligence, racial bias, and more. I hope you enjoy the journey.

TECH CAN BE UNFORGIVING

Sometimes technology can seem cruel, hard to learn and often downright unforgiving And nobody likes being disrespected. So, let us get familiar with the cyber world we live in.

The contents of this book will be useful to anybody whose life is intimately touched by cyber technologies -- basically everyone on the planet. It does not matter which country or continent you are on, technology will play an important part in your life.

THE LIVES WE LEAD

We live two different lives. We live a life in the physical world, and we live another life in the digital world. Our physical and digital worlds, while separate, frequently interact. That interaction is near seamless.

The interconnectedness of both worlds is illustrated when we physically engage with our phones connecting to members of our social community, browse digital news feeds, and remotely purchase goods from local marketplaces through self-guided apps. These digital world activities have physical world associations and implications. Creating a unique world, we are all figuring out together.

WE NEED SOMETHING SPECIAL

Cyber technologies are all around us, and modern life requires us to engage with them. Almost like an invasive plant species, they can seem intrusive to our physical lives. Technology can be draining, distracting, and hard to incorporate in daily life. Yet, technology can be energizing, connecting, and provide ease to our daily life. It is the duality of the effect's technology can have, so we need something special to help. We need a different mindset; a mindset willing to change with the ever-changing technology landscape.

GETTING THE CYBER MINDSET

GETTING THE MINDSET

Born in the 1990s, I do not remember my first computer, when I got my first cell phone, or when I first connected to the internet. However, I do remember being in complete disbelief of everything technology and what the internet allowed me to do -- from chatting with friends around the world, meeting strangers willing to drive me wherever I wanted to go, and finding groups that share my niche hobbies.

Technology has truly changed not just my life, but the lives of everyone living and the future generations to come.

However, nobody ever taught me how to use technology like we teach driving skills. I am sure many people reading this will feel the same. Sure, I knew what a power button did, but I was not familiar with of digital organization, keeping myself safe on the internet, or even how to purchase technology that would positively impact my daily life. These are subjects and skills which are now more important than ever.

Understanding technology, the internet and cyberspace as a whole -- where many people now spend most of their time -- requires a different type of mindset. Some old ways of thinking are still effective in navigating this world, but it is time to cultivate some new ways of thinking.

FOUNDATION OF THE MINDSET

You already have the foundation to support this new mindset. If you own a mobile phone or a computer and you want to know how to use it, then you already have the mindset of a cyber guru.

MINDSET BASED ON CURIOSITY

We are amazing creatures! We are human after all, and humans by default are equipped for change. We are always ready for the next mind-blowing innovation or big breakthrough. At the core of what fuels our readiness for change is simple, CURIOSITY.

Curiosity allows our mind to be ready for the unexpected. While humans are innately curious, we might not be inclined to explore this nature; even though an interest is present. But the call to be curious is there.

If you have ever wondered how to prepare for the next mind-blowing breakthrough or feel as though technology is moving faster than you can keep up with, remember that at your core you are a curious human being. Remember the driving factor that took humans to the moon, helped us discover the wheel and master fire. It was all CURIOSITY.

OVERWHELMING CURIOSITY

Although we possess an innate curiosity, learning anything new can feel overwhelming and that is okay. I have been overwhelmed with learning about technology before. There can often be too much information to consume and comprehend, but a little bit of curiosity goes a long way.

With a bit of curiosity, you can become a master of any craft, subject, or mindset. You can conquer any terrain. You can do anything. It is your superpower, but you can still feel lost in information overload.

To not feel lost in the flood of information, all it takes is a smidgen of curiosity. Curiosity to experiment, read about and even question the usefulness of new technologies. The more you investigate, the more you feed your curiosity monster, the less intimidating and overwhelming advancements in science, technology, engineering, and math will feel.

STAY CURIOUS; GROWTH MINDSET

We all grow older and most of us become less flexible as we age. To better understand the world, and that of cyber, we need to make sure our mindset does not become stagnant.

I have missed many experiences because of a stubbornness to socialize, explore or get involved. This is unfortunate because growth rarely happens in isolation.

The solution, to the problem of how to stay mentally flexible, is as simple as talking with others. Communicate consistently with people drastically different than you -- people younger, older, richer, poorer, and more educated than you; that means creating a dynamic social circle.

Make it a priority to capitalize on opportunities to develop your social circle. It is a must-have in the world of technology we live in today. We are no longer burdened by the limits of physical distance. Our only limits now are being steadfast in old ways, perspective, and comprehension. All of which can be overcome.

DYNAMIC SOCIAL CIRCLES

I believe we reflect traits of those closest to us: friends, family, romantic partners, and coworkers. If most of us looked closely at our social circles, we might realize a sad truth fairly quickly; our social circles have little to no variety or diversity.

We can learn more from people of different races, countries, cultures, religions, sexualities, professions, and age than by surrounding ourselves with people who are the same as us or that we naturally relate to. Otherwise, we become stuck in a repetitive cycle where nothing new comes in and nothing new goes out.

Social circles without variety do not encourage growth. Without growth curiosity dies and exploration never begins. New ideas often don't appear and if they do, they are quickly rejected and dismissed without hesitation. The status quo does not change, and our mindsets stagnate.

To learn, explore, and grow -- to be curious -- we must not remain average. We must diversify. We must have dynamic social circles. We must seek beyond what we internally hold true because that truth could be a falsehood. We must remain adventurous in our endeavors, especially to fully grasp the world of computer technologies. We must embrace our curious nature.

STAYING CURIOUS; EXPLORE

Here is a list, in no particular order, to help us stay curious and keep exploring on our journey of learning what is cyber.

- Stay curious, especially when faced with confusion.
- Keep your social circles diverse, find masters to teach you and be willing to teach beginners what you learned.
- Remain childlike in a sea of knowledge and wisdom; staying curious when all seems overwhelming.
- It is okay to not know something, the sin is not wanting to learn.
- Ask questions. Ask LOTS of questions. Especially ask the 'dumb' questions because no question is ever truly dumb.
- Life is not nice and neat. Focus on the messy infinite possibilities.
- Learn the way you learn, whether it is by reading, watching, doing, hearing, or a mixture of those. Learn how you as an individual best gain knowledge.

THE BASICS

STARTING WITH THE BASICS

In order to walk you first must learn to crawl. It is an old adage, but quite true. And it is especially applicable when we are learning anything computer related. So, let us ask ourselves an important question: growing up, did anybody teach me how to interact with computer technologies, how they fit into my life, and how to use them?

Where do you fall on the spectrum of yes, no, and somewhere in between? If you answered yes, then you are already well ahead of many, and the following chapters will firmly cement your base knowledge. If you are anywhere else on the spectrum of yes, no, and somewhere in between, then the following page will be a good introduction to computing basics.

BORING, MAYBE

The basics often seem boring and unnecessary. However, ask any pro athlete and they will often speak highly on learning the fundamentals and their importance. The same can be said for cyber professionals and computing basics.

We have to start with the fundamentals in order to fully understand the world of technology. We are here trying to develop the tools and mindset to stay curious and explore this amazing world of cyber.

First, we are going to learn how to count and understand numbers the way our computers do. It is essential, as you will see in the following page.

ARE YOU COUNTING RIGHT

Everybody counts. It is one of the most fundamental skills needed to survive in life. However, most people actually count incorrectly; or rather conceptually misunderstand the most commonly used numbering system in existence: decimal. Nothing explains this more than the fun fact that the number 10 'doesn't exist'.

Reflexively, we might think to ourselves, 'but I can count to 10'; and that is true. We can count to 10 and beyond. But the number still 'doesn't exist'. At least, not in the decimal system.

WHAT'S DECIMAL

Decimal is a base ten (base 10) numbering system. Which means the system only contains ten numbers: 0, 1, 2, 3, 4, 5, 6, 7, 8 and 9. A numbering system is a method of using characters to represent numbers. The characters most related to the decimal number system is called Arabic numerals. If you are wondering what Arabic numerals are, you probably already know them. They are 0, 1, 2, 3, 4, 5, 6, 7, 8 and 9.

BINARY

We've all probably heard the concept of binary in passing conversation. We might have even wondered what it is. Here are the details.

Binary is a numbering system, similar to the decimal system. It is a base two (base 2) numbering system. Meaning it is a system made up of only two numbers: 0 and 1.

HEXADECIMAL

Much like binary and decimal, hexadecimal is another numbering system. It is a base sixteen (base 16) numbering system. Hexadecimal is a little different than binary and decimal though. This numbering system includes: 0, 1, 2, 3, 4, 5, 6, 7, 8 and 9. After 9 the notation changes a little to include letters: A, B, C, D, E and F. Since the number 10 'doesn't exist'. A, B, C, D, E and F respectively represent 10, 11, 12, 13, 14 and 15.

WHERE'S THE BASE

There are so many different numbering systems. The most commonly used include base 2, base 8, base 10, and base 16.

Seeing a theme? Or are we asking ourselves this question, where is the base number when counting in these systems? Where is the 2? Where is the 10? Where is the 16? The answer is simple if we have not already figured it out. Much like in base 10, the number 10 'doesn't exist'. In base 2, 8, and 16, the number or character representation of 2, 8, and 16 'doesn't exist'. The base number just represents how many characters are in the particular numbering system.

Although the number systems are referred to as 'base 2, 8, etc.', the numbers themselves (2, 8, 10 and 16) are not physically a part of the system. For example, base 2 is made up of two numbers -- 0 and 1. If the number 2 was included in base 2, it would not be referred to as base 2 anymore because there would be three numbers in the system (0, 1 and 2). If the number 2 were added, it would change the base 2 numbering system to a base 3. You may also wonder, 'can't a base 2 system be made up of the numbers 1 and 2?'. The short answer is no because the numbering systems all start with the lowest known number (0) and progresses in numerical order, 0 then 1 and so on.

Let us look at the octal numbering system to drive the point home. Octal is the base 8 numbering system mentioned earlier, where the number 8 'doesn't exist'. So, the only numbers that matter to us in the octal system are: 0, 1, 2, 3, 4, 5, 6 and 7.

CAN BASES BE SIMPLER

There is a simple trick to remembering how to conceptualize our numbering systems.

Remember analog odometers, those devices in cars used to track distance traveled? Those old things, they have been made digital now, but served as a great way to visualize our decimal numbering system. After the number 0009, it always rolls over to a 0000; while creating a 1 in the next numbering place 0010. That is our decimal numbering system at work and in action.

Our numbering systems are important because computers are all based on these systems, and they use these systems. The world of computer technology is shaped by numbers. Every image we see on our computer screens, every phone call made with a cellphone, and every post we share on social media is all built on the foundation of numbers. Starting with 1s and 0s, which represent On and Off.

0 0 0 9 → 0 0 9 9 → 0 0 1 0

THE CONNECTED WORLD

The digital world we live in appears to be invisible, however, this is simply not true. The digital world is connected by wires and entangled with numbers -- wires are connecting us from continent to continent, home to home and person to person; while the numbers of this connected world breathe life into everything we do. And that is the magic of computer networking. It seems like a magical world that appears invisible to only a select few who can see it. But can we all see it? The answer is simple. Yes.

IN THIS MAGIC WORLD

In this magical world of computer networking, only a select few seem to be capable of making sense of it all. However, they only see the strings in the web connecting us. If we believe, explore and study, we can all see the strings.

IT'S NOT MAGIC

Much like homes we live in, every connected device has an address. Much like the mail we receive at our home address, every connected device can receive mail. Much like how we can open and read mail, every connected device can open up and use/read the information it receives.

This magical world of the digital space is not magic. It is very real and tangible.

We are not connected by an invisible or magical force -- a force some might believe to be beyond comprehension. We are connected by something real. Something we see every day. Something which is easily understandable. So, let us understand better and see that the digitally connected world is not magic.

HOW COMPUTERS TALK

Computers can talk to each other. They talk a lot, and they talk often. Some computers speak and wait for a response, so as not to be rude. Some computers talk frequently, never waiting for a response. They do not really care about politeness, they never give their fellow computers a chance to speak.

These two contrasting forms of speaking are called the Transmission Control Protocol and the User Datagram Protocol: or TCP and UDP, respectively, for short. TCP is connection-based. It wants to make sure the other computer gets a chance to communicate. UDP does not care about anything the other computer has to say, it just cares about getting its own point across. It is so RUDE.

The most important aspect for computer users of TCP and UDP is that both of them are networking protocols.

NETWORKING PROTOCOLS

Much like how humans use language to speak and communicate, computers do the same. One-way computers communicate is through something called networking protocols; like TCP and UDP. A networking protocol is an agreed standard of how information of a certain type should be passed from one point to the next. You can call them digital languages.

There are a lot of different networking protocols used by our digital devices. Let us focus on the one most noticeable to us, the Internet Protocol or IP.

PROGRAMMING LANGUAGES

Before we jump into the Internet Protocol and some other related topics, let us have a chat about programming languages. Programming Languages are very similar to networking protocols. In a sense, they are both ways for our computers to communicate. What is the difference? We use programming languages to communicate what we need our computers to do. Programming languages make up the visual and functionality of websites, phone applications (apps), and other software. While networking protocols are used to carry the data of those programs from computer to computer.

There are numerous programming languages and new ones are being invented every day, so I have always compared them to dialects.

HOW ARE THEY DIALECTS

A dialect is a certain way a group of people speak. All programming languages, at their core, communicate the same things. The how and interpretation of communication is just different.

A real-world example of this is easy to create. Imagine two English speakers having a talk, but one is American and the other is British. They, at the core, are speaking the same language. But the how and interpretation of certain words and phrasing are slightly different.

Now on to the Internet Protocol.

THE INTERNET PROTOCOL

Every internet connected device has an IP address and IP simply stands for Internet Protocol. Internet Protocol is a networking protocol and we already know networking protocols are how computers speak to one another. So, we are already ahead of the curve.

Our toolkit is starting to fill out a little bit. So, yay for mini wins. However, there is still a bit more to learn about the Internet Protocol, and the importance it plays in our lives.

SHORT HISTORY OF IP ADDRESSES

IP addresses are running out. There are not enough of them. Get yours while supplies last. Those first three sentences were true at some point. IP addresses were actually running out, meaning any new device wanting to connect to the internet would not be able to, because the addressing space and resources were so limited.

Now we have two main different versions of the Internet Protocol. One is called IPv4 and the other is IPv6. The 'v' in both IPv4 and IPv6 simply stands for version.

MORE ABOUT IPv4

IPv4, made in the 1980s, was the original stable way for computers to talk across the net allowing us to share cat pictures.

Most of us might even recognize an IPv4 address if we saw it. Because it looks something like this: 127.0.0.1.

The issue with IPv4 was a small pool of assignable addresses. IPv4 only had 4,294,967,296 assignable addresses or 2^{32} or 32 bits worth of space (we will talk about what a 'bit' is later).

Now this might seem like a lot of space, until we consider some interesting facts. At the time of writing this, about every household in the United States of America (USA) alone had about 10 internet-connected devices. There were roughly, at the time of writing this, 119 million households in the USA. Which roughly means 1.2 billion assignable addresses would be used by the USA on personal devices. This does not factor in past or future devices. See how easy it was to run out?

There were some other solutions implemented to help create more usable IP addresses. But the real hero is IPv6.

MORE ABOUT IPv6

Developed in the 1990s, IPv6 is the new model. It fixes the issues of IPv4 and even provides a bit of security built into the protocol itself. However, it is not fully adopted yet. The slow adoption of IPv6 is due to older technologies and infrastructure being heavily connected to IPv4. There are many hurdles to migrating over to IPv6. Cost and legacy systems are just two concerns. But one major roadblock is no one sees a direct need for migration. Currently. The saying if it ain't broke don't fix it comes to mind. However, what most network administrators and engineers know and understand is one day it will break. So, we should probably start fixing it now.

With IPv6, we should not run out of IP addresses anytime soon. Let us brace ourselves, because IPv6 sports a hefty 340,282,366,920,938,463,463,374,607,431,768,211,456 IP addresses. That is 2^{128} or 128 bits worth of IPs.

Hopefully, we are all alike here and instantly thought 'wow, that's a lot'. And we are right. It is a large pool of IP addresses. The primary goal of IPv6 is making sure we never run out of IPs addresses anytime soon.

We also might recognize what an IPv6 address looks like right now. If not, it is probably a good idea to get familiar with them. Because they look drastically different from an IPv4 address. IPv6 addresses look like this:

3FFF:1F00:4545:3FFF:200F:F8FF:FE21:FFCF

STARTING TO BRING IT TOGETHER

Earlier we learned about different numbering systems; the main three being binary, decimal, and hexadecimal. These numbering systems can be simplified into bit format.

A 'bit' in computer speak actually stands for Binary Digit. A 'bit' is the most basic and smallest unit of data in computer science. A bit is just a single binary digit of 1 or 0.

A binary digit or bit is used in binary. We might be thinking, 'yea that's super obvious'. But what is the length of a binary number?

The length of a binary number is the amount of 1s and 0s it has. For example, 01010011 has a length of 8 bits.

BIT, NIBBLE, BYTE

When discussing data, there are some common terms we use. They are bit, nibble, and byte. Most of us are familiar with a byte because we often see it prefaced by kilo, mega, giga, or tera. But a byte's size is actually 8 bits or 2 nibbles. A nibble is 4 bits in length.

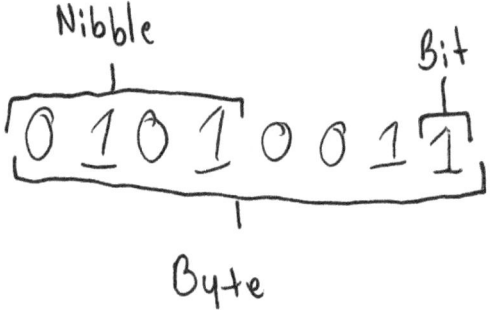

QUICK ABBREVIATIONS & VALUES

Here is a quick reference guide to how we would abbreviate those combinations.

Kilo + Byte is KB and equals 1,000 bytes

Mega + Byte is MB and equals 1,000 kilobytes

Giga + Byte is GB and equals 1,000 megabytes

Tera + Byte is TB and equals 1,000 terabytes

There also exist values higher than terabytes. They are petabyte, exabyte, zettabyte, and yottabyte. But for our daily lives, we only really need to remember kilobyte, megabyte, gigabyte, and terabyte.

A LITTLE MORE ABOUT IPv4

Now, let us really start to bring everything together we learned about IPs and data. Here are some quick facts about the structure of an IPv4 address:

- 127.0.0.1
    - This is an example of dotted-decimal notation; which is used for IPv4.
- An IPv4 address is split into 4 sections.
- Those 4 sections are called Octets.
- An Octet is 8 bits or 1 byte of data.
- The entirety of an IPv4 address is 32 bits or 4 bytes of data.

A LITTLE MORE ABOUT IPv6

Here are the quick facts about the structure of an IPv6 address:

- 3FFF:1F00:4545:3FFF:200F:F8FF:FE21:FFCF
  - This is an example of hexadecimal notation.
- An IPv6 address is split into 8 sections.
- Those 8 sections are called Hextets.
- A hextet is 16 bits or 2 bytes of data.
- The entirety of an IPv6 address is 128 bits or 16 bytes of data.

MAGIC OF THE NUMBER 2

For most tech geeks, 2 is a really magical number. Every electrical current running through a piece of hardware or represented in software starts with 2 distinct signals, on or off, true, or false, 1 or 0. That is binary, existence in one of two states. Binary itself is representative of a base 2 numbering system. We count in this system via the 'power of 2'. We can count every number in existence with a medley of 0 and 1s side by side, which is just 2 numbers. This is the magic of 2.

THE REAL BASE 2

Earlier we chatted about some of the basics of base 2 and other numbering systems. With this knowledge, we are starting to embrace our number nerd. Hopefully, we are beginning to comprehend the importance of base 2 in our lives. But we are still missing a vital piece. We must be able to easily count in base 2.

Much like we can count in base 10 and know the numbers it uses (0, 1, 2, 3, 4, 5, 6, 7, 8 and 9), it's essential to be able to count in base 2 and know the numbers it uses (0 and 1).

Counting in base 2 is easy. If we want to know what comes next in base 2, then we simply double the number we are on. I.e., 1, 2, 4, 8, 16, 32, 64, 128, 256, 512, 1024 and so on. Why is knowing all of this so important? Because everything is data.

$$\underbrace{\frac{0}{128} \ \frac{0}{64} \ \frac{0}{32} \ \frac{1}{16} \ \frac{0}{8} \ \frac{0}{4} \ \frac{0}{2} \ \frac{0}{1}} = 16$$

1 byte = 8 bits

$$\underbrace{\frac{1}{128} \ \frac{1}{64} \ \frac{1}{32} \ \frac{1}{16} \ \frac{1}{8} \ \frac{1}{4} \ \frac{1}{2} \ \frac{1}{1}} = 255$$

1 byte = 8 bits

BRINGING IT TOGETHER VISUALLY

Below is an IPv4 address breakdown:

```
  1st Octet   2nd Octet   3rd Octet   4th Octet
    192        168          16         254
  11000000.10101000.00010000.11111110
  └──┬───┘
1 bytes = 8 bits
  └─────────────────┬──────────────────┘
              32 bits (4 bytes)
```

Below is an IPv6 address breakdown:

```
  2001:FCB5:0000:0000:04EA:00FF:5DA3:005B  — IPv6 Address
   └┬┘                 ↑↑↑
  Hextets = 16 Bits   0000 0100 1110 1010  — Binary
                      └────────┬────────┘
                             4 Bits
           └──────────────────┬──────────────────┘
                          128 Bits
```

JAE TYLER | 56

WHY IS ANY OF THIS IMPORTANT?

IP addresses, networking protocols, our numbering systems, and programming languages all communicate data from source to destination. But the practicality of knowing this information is not solely tied to working within the information technology fields. We do not have to be a network administrator or software engineer to get some utility out of this knowledge. It is a building block for the curious mind to explore beyond it.

We could learn how to program for personal use. Programming languages can be used to automate daily tasks. Like starting the coffee maker, responding to emails, or even creating a grocery list.

Knowing about networking protocols and IPs allows us to communicate better with our IT support teams at work. It can be empowering as well. Because now we really grasp the underlying mechanism of our home networks and personal devices.

The world of computers is not magic. It is just data. Data is just concepts about numbers.

DATA, DATA, DATA

Everything is data and data is everywhere. Our physical world we live, feel, and breathe in is nothing more than data. We quantify and qualify that data internally as humans. We process through our senses of touch, taste, hearing, vision, and smell. We decide how to use information data provides to us, because we are in essence computers ourselves. We are curiosity driven data absorbers and generators.

Now we live in a world where biological data can be easily converted into digital. Knowing what our digital avatar's composition might not play directly into our daily lives. But it should give perspective. A perspective of -- we might not know how something works in great detail, but at least we can understand the components. A great parallel would be a car mechanic versus a car driver. The driver understands the basic anatomy and purpose of a car and can communicate issues to a mechanic.

Which really demonstrates; the way we interpret and comprehend data is just as important as the information to be collected. Because without the ability to communicate issues, can we really resolve problems? The ability to communicate issues properly means we are one step closer to feeling comfortable in a world of data and getting used to our digital life. Our life connected by technology.

INTERNET OF THINGS

INTERNET OF THINGS

The Internet of Things or IoT is composed of every device connected to the internet. The internet is best described as a global network of computers. So, whether it is our smartphone, TV, car, or planter pot sending us notifications that our plants are thirsty and need sun, these devices and more are included in the IoT. The IoT seeks to bring connectivity to our once disconnected and disjointed world. Because of this new connectivity, everything seems to be getting an intelligence boost. We are taking dumb devices and making them smart devices.

IoT AND ITS GOAL

The IoT can be thought of as devices talking to one another, with the main goal being the creation of a digital ecosystem where all devices communicate their wants and needs to the relevant devices to have those wants and needs fulfilled. There are two great examples of this we might see today and so many more either exist or will exist.

Example 1: You have security cameras and a smartphone. The cameras are mounted in every room of your house. They are set up to trigger if someone enters the room without a recognized smartphone. They are disarmed if you enter the room with your smartphone. In this example, our phone and security cameras are actively communicating via internet connection or some other type of network. They are aware of each other; having a conversation and communicating about the needs and requirements of each related device.

Example 2: All phones traveling on roads send location information to the Global Positioning System (GPS). With that information, global mapping systems, like Google maps, attempt to optimize routes so that every commuter has a smoother, faster, and less stressful drive.

WHAT DATA BECOMES

Our devices talk to each other. We already learned how they communicate through protocols. They are transmitting data and our personal devices are constantly giving out data on us. What we buy, what we like, what we watch and where we have been are all known by someone or something. What is all this information used for? Crafting an avatar of sorts.

LIFE ON THE INTERNET

Ever used the internet? The answer is probably the same as mine: yes. If you have not that is perfectly okay.

We live in a hyperconnected, super integrated and massively social society. It might not even come as a surprise to learn that we all have digital avatars. Digital avatars represent a stitched-together concept of our daily lives on the internet.

Our digital avatars represent our physical self and life. They are interwoven concepts of our daily lives on the internet, much like Frankenstein's monster cobbled-together body. It is a programmed algorithms' attempt to make sense of all the data we actively and passively provide. And it is almost scary how accurately our digital avatars describe who we are, where we have been and what we are interested in.

OUR INTERNET AVATARS

Our internet avatars do not have arms, legs, or a brain, but they represent us. They are composed of data we give freely to the gatekeepers of the land of the internet and memes – meaning websites, apps, and services provides (Netflix, Facebook, etc.). Our avatars live without us, but they cannot live without us. Everything they are is us. But they are not everything that we are.

GATEKEEPERS OF THE INTERNET

In the world of the IoTs, there are some gatekeepers of the internet, regularly referred to as Internet Service Providers or ISPs. Without an ISP, we would not be able to access, interact, and contribute to the rich world of the internet. We might even only have access to certain types of ISPs, depending on where we are located in the world. Please know not all ISPs internet connections are created equal.

TYPES OF ISPS

There are about six types of internet services an ISP can provide to a customer and not every ISP offers all the types of services. Each service has advantages and disadvantages. Here is the breakdown:

1. Dial-up
    a. Slowest and most outdated service.
    b. Great in areas without faster options but have telephone lines.
2. Satellite
    a. Popular in remote and rural areas without access to other types of ISPs.
    b. Much faster than dial-up but much slower than other services.
    c. Hard caps on data use, especially during peak hours.
3. Cable Modem
    a. Faster speeds than dial-up or satellite.
    b. Allows for streaming video and gaming.
    c. Typically, available from most cable companies.

4. Digital Subscriber Line (DSL)
   a. High speed internet over telephone lines.
   b. Allows for streaming video and gaming.
   c. Typically, available from most telephone companies.
5. Mobile Broadband
   a. Internet through cell phones often referred to as a wireless internet service.
   b. Allows for streaming of video, but gaming not so much.
   c. Service provided through most cell phone companies.
6. Fiber Internet
   a. The fastest type of internet service out now.
   b. Allows for effortless streaming of multiple videos and gaming.
   c. Not currently available everywhere.

INSIDE A COMPUTER

WHAT IS A COMPUTER?

We have gone over a lot things in this journey to become more comfortable with technology and everything it offers. Yet, we have not covered what is a computer.

By definition, a computer is a programmable, usually electronic device that can store, retrieve, and process data. By this definition, computers can include your desktop, laptop, phone, tablet, smart TV, smart washer/dryer, and more. In reality, pretty much anything can be a computer.

Here is an enjoyable thought. Humans could be considered a computer. We are programmable. We can store, process, and retrieve information for later use. Humans are computers.

THE ANATOMY OF A COMPUTER

Have you ever looked at a computer and wondered, 'what the heck is in that thing?'. The answer to that question is really simple. So, let us talk about the anatomy of a computer.

In a computer's most basic form, it needs these three things: a way to process information, a way to store information, and a way to power those processes. There are 5 components in a desktop, laptop, or handheld device doing everything we require. These necessary components are:

- Central Processing Unit
- Random Access Memory (RAM)
- Read-Only Memory (ROM)
- Motherboard
- Power supply

Each performs a simple task. Helping create a fully functioning computer.

CENTRAL PROCESSING UNIT

The Central Processing Unit or CPU can be called a central processor. A CPU handles all the computational work of our computers. We can think of them like this; any information we give a computer the CPU takes, processes, and sends everything where it needs to be.

Processors are all about speed. Speed, for CPUs, is measured in hertz (Hz). The higher the hertz, the faster the CPU.

The CPU does so much. When we watch videos, photos, or memes, this is often handled by our trusty CPU. However, there does exist a Graphic Processing Unit or GPU. A GPU allows us to enjoy the highest resolution videos and videogames.

GPUs are specialized processors. That work to help alleviate the burden of processing visual data CPUs might struggle with. GPUs can even be used for strange and odd tasks, because they are so specialized and powerful. For example, when mining cryptocurrencies, GPUs are typically used over the generalized CPU. Every computer still needs a CPU to run, while GPUs are just for fun and games. Pun intended.

TWO TYPES OF MEMORY

The CPU may run the show, but without memory, our computers cannot properly function. There are two types of memory: Read-Only Memory (ROM) and Random-Access Memory (RAM). Both play a part in keeping our computers functioning.

READ-ONLY MEMORY

Read-Only Memory, also known as ROM, is true to its name. Computers can only read information from this type of memory. The contents of ROM cannot be changed, altered, or reprogrammed. It is non-volatile, kind of like this book. We can read it all day long, but we can never rewrite its contents. We would never really want to change anything about ROM, because this is the place our computers go to for information on how to power up, know basic input/output systems (BIOS), and basic data management. Without ROM, our computers would not know what to do after we hit the power button and the lights come on.

RANDOM-ACCESS MEMORY

Random-Access Memory, also known as RAM, is volatile. RAM is read and write memory. Think of it like a journal. We can read journals; we can write in journals; we can even erase all the information inside the journal, leaving the pages blank to write in them again.

Within the conceptual framework of RAM, there exist hard drives. Hard drives are considered Random-Access Offline Storage (RAOS). Meaning data can be saved and stored while our computer is powered down.

All data we create on our computer is stored on a hard drive. For example, every document for work, school assignment turned in late, or vacation photos of our family sits on a hard drive, so we can revisit those memories at a later time.

There are many different types of hard drives. You might already know a few of them. But there are a number of different devices that also count as a Random-Access Offline Storage. They are:

- Hard Disk Drive (HDD)
- Solid State Drive (SSD)
- Universal Serial Bus Drive (USB)

MOTHERBOARD

We could think of the motherboard as the nervous system of the computer. Motherboards provide all the connections inside of our computers. A motherboard houses the RAM, CPU, and hard driver connections. Motherboards, also, provide connection points for: HDMI cables, a mouse, a keyboard, printers, scanners, game controllers, and much more.

POWER SUPPLY

All the components need power to run. The power supply provides the energy for our computers. It is the simplest piece, but essential for getting our computer to work. A power supply can be a battery or a plug connected to an electrical outlet.

BUILD OR BUY

The components of computers are easy enough to understand. Which brings two questions to mind: is it easy to build a computer and should you build or buy?

It is quite easy to build a computer. If this were a more detailed book, I would even go step by step over the process. But do not worry. There are numerous online guides and tutorials going over the process. These resources make building a modern computer simple and stress free as possible. But the question of should you build or buy requires a deeper discussion.

In the personal computer (PC) world, there is a long-standing debate on whether to build or buy. If you are in the market for a computer, then you might often go to an electronics store and just buy what the salesperson recommends. But next time you will be equipped with a bit of computer knowledge: your new powerful tool to get your money's worth.

Desktop computers are a unique category of electronic devices. Unlike phones, laptops, or TVs, you can easily build a desktop or just as easily buy one prebuilt, which leaves many different reasons to either build or buy a computer. But there are only two reasons I keep in mind when recommending building or buying to anyone. I simply ask them, 'Are you a tactical person or do you prefer plug and play?'

If you are a tactical person, then take the chance of building your own PC. You might enjoy the experience and you get to brag about having built your own custom rig (PC) to your friends/family while saving a bit of money.

If you are more plug and play, meaning you like keeping setup minimal, then do a bit of research and get yourself a PC that fits your lifestyle. Both are honestly good options, it just depends on your personality.

After building or buying, it would be nice to know a few keyboard shortcuts, right?

KEYBOARD SHORTCUTS

Here is a fun fact; computers are designed in such a way, that only a keyboard is needed to operate them. Making the mouse and any other input device an accessory and not really a necessity.

Let us go over some useful keyboard shortcuts. Because keyboard shortcuts can make our lives easier. They are primarily PC centric but there are a few Mac equivalents. These will be split PC native and Web Browser based shortcuts.

PC NATIVE SHORTCUTS

PC Native Shortcuts work everywhere:

- Ctrl+Z: Undo
  - It will rollback your last action on your computer. This means everything from typing to deleting a file (long as it exists in the recycling bin).
- Ctrl+W: Close
  - This will close out of windows, web browser tabs, and other similar programs.
- Ctrl+A: Select all
  - This command allows for selection of all the text in a document or files in a folder.
- Alt+Tab: Switch apps
  - When you are using multiple applications. The ability to switch between apps can save a bunch of time and time is money.
- Alt+F4: Close apps
  - I am throwing this in the list because I saw so many people trolling. The troll would go something like this:

- Victim: How do I do <thing>?
- Troll: Hit Alt+F4 and it will do it
- Victim: *types Alt+F4 then application closes, and all their work is lost*
- Victim: ...what the f*ck???
- Victim: Technology sucks!!!

WEB BROWSER SHORTCUTS

These are some useful web browser keyboard shortcuts. Do not be too concerned with what web browser you are using. These are universal. This is not a comprehensive list.

- Ctrl+T: New tab
    - This command opens a new tab.
- Ctrl+W: Close tab
    - This command closes the current tab.
- CTRL+N: Open new window
    - This command opens a new browser window.
- Ctrl+Shift+T: Open previously tab
    - If you ever close out of a tab accidentally, then this is the command for you. This command opens the previously closed tab.
    - It also has a lesser known function. If you ever close out of a web browser losing all your tabs, you can then open your browser again and just type Ctrl+Shift+T. And boom all those lost tabs are back.
- F5 or CTRL+R: Refresh

- This command reloads the current tab. This helps out if something does not load properly on the website you are visiting.
- CTRL+H: Open history
  - This command opens browser history. Just in case you ever need to check what site you have been to.

COMPUTERS CAN HAVE PROBLEMS

Sometimes computers break or malfunction. Much like everything else in life, computers are not perfect and figuring out the issue or fix can be frustrating. At times, the fix is as simple as turning it off then on again. This is called power cycling, but other issues can require more detailed solutions.

TROUBLESHOOT IT

When power cycling does not work, we have to start using the skills and knowledge gained in our toolkit. We have to troubleshoot.

Troubleshooting, a lot of times, requires following exact and detailed steps, normally tailored to a specific device. But there are some basics of troubleshooting that should be followed in any circumstances. Note these are just basic things anyone can do for their personal devices.

- If the device is not powering on, check and make sure the cables, battery, or power source are firmly attached to the device.

- If a program, software, or app is not connecting, make sure it is updated to the latest version.

- Watch for and read error codes, then go and look them up on the internet. The issue and solution are often well documented.

- Identify the problem in simple terms, then research the issue.

- When all else fails seek trusted professional help.

HAPPY COMPUTERS

A clean computer is a happy computer. Having a dirty or dusty computer can affect its functionality because computers need proper airflow to function. Computers can overheat and become damaged without proper airflow, much like us on a hot summer day.

A computer that is properly and regularly cleaned will last longer and work reliably. This keeps us productive and saves us thousands of dollars in the long run, because we are not frequently replacing malfunctioning computers.

SHOULD WE CLEAN OUR DEVICES

Germs are everywhere and often in places we might never think about. The same amount of times we clean our kitchen, bathrooms, or ourselves, we should clean our digital devices -- like everything else in life, our electronics can get dirty, dusty, and carry germs.

PHONES NEED CLEANING TOO

Unclean phones, phone cases, and other accessories can also carry bacteria, germs, and viruses. It makes a lot of sense, right? We take our phones everywhere. They are an ingrained part of our lives almost like extensions of ourselves.

Let us think about all the places we take our phones. They go with us to the bathroom, to bed, out in the garden, the barbershop, work, the list goes on and on. We take our phones everywhere. When we touch something more than likely we are touching our phones right after at some point. Which could cause bacteria and germs to spread. Meaning we should really start to clean our phones and develop the habit. Using a disinfectant wipe is a simple and easy way to clean our phone screens and cases. Cleaning a desktop, however, is a bit more involed.

GETTING OUR TOOLS READY

Before we get started cleaning our desktops, first, we will need some cleaning tools. We will need everything below:

- A Screwdriver
- Can have Compressed Air
- An Unused Paint Brush

These are some optional items:

- A Face Mask
- An Anti-Static Wrist Strap
- A Small Handheld Vacuum

HOW TO CLEAN OUR DESKTOP

Here are some basic steps to clean our desktops:

1. Shutdown the computer, unplug from the wall outlet, and hold the power button down for about 30 seconds.
2. Unhook anything connected to the computer. This could include but not limited to: keyboard, mouse, printer, webcams, monitors, ethernet cable, power cable.
3. Take the computer to a clean, clear, and dry location.
4. Attach your anti-static wrist strap if you have one.
5. Use a screwdriver to open the computer case.
    a. There will typically be a set of screws on the back. Those screws keep the side panel attached. Unscrew them and the side panel should come right off.
6. Use the can of compressed air to blow dust out of the computer.
    a. Put on a face mask if you have one.
    b. Do not touch the inside of the computer. Our bodies can carry

electricity, which could lead to a static shock damaging the internal pieces of the computer.

   c. Wear an anti-static wrist strap if you have one.

   d. Use short bursts of air.

   e. Do not allow the can to leak onto the computer parts.

7. Make sure all fans are clean and free of dust.

8. Check for any dust hiding in corners or hard-to-reach places.

   a. This is where the paintbrush is handy.

9. Lastly, vacuum up all the dust, close up the computer, and enjoy a happy healthy dust free PC.

A few words of warning. Computer parts are delicate. We should never force or be rough with them. Be gentle and be patient.

CLEANING OUR LAPTOP

Just like cleaning a desktop, it is equally important to clean our laptops. But not to worry. Cleaning a laptop is very similar to cleaning a desktop. With the exception that we will need a special set of screwdrivers, called precision screwdrivers. We do not have to worry about breaking the bank. They are normally pretty cheap to purchase.

Now, we just follow the steps for cleaning our desktop. For additional help, the internet is our friend. There are tons of videos of people opening up and cleaning lots of different makes and models of laptops, meaning we do not have to go in blind.

CLEANING OUR SCREENS

Our world is littered with screens -- dusty, dusty, dirty, fingerprint-riddled, and smudged screens. We've all probably had this thought of 'hey what is...how did that get there' about screens. We should properly clean them like everything else.

Cleaning our screens is actually really easy. The only tools needed is a microfiber cloth and maybe a mild soap solution which is part water and soap.

Follow these steps:
1. Turn off the screen.
2. Dust/wipe off the back of the device.
3. Wipe screen with dry microfiber cloth.
    a. Use light pressure for tough spots.
    b. For really tough spots, apply a very light amount of mild soap solution to cloth.
    c. Never apply solutions directly to the screen.
4. Let completely dry if a mild soap solution was used.
5. Enjoy a clean screen.

CLEANING OUR WORKSPACES

Everybody's mindscape works a bit differently, so there is no right or wrong way to have a workspace laid out. If we feel more efficient with a minimalistic workspace, then let us keep it minimalist. If we feel more efficient with knick-knacks and trinkets everywhere, then keep those items which inspire us. However, we must keep our workspaces clean.

Having a clean workspace does not mean having only a few items on your desk or keeping them locked away in cabinets. Having a clean workspace means limiting trash, dust, dirt, and items we find personally unnecessary. Having a clean workspace means ridding ourselves of items that do not inspire or drive productivity for our chosen task. Having a clean workspace gives freedom to our unique mindscapes. Having a clean workspace allows our digital devices to be happy in the space they live in. How is this possible? Simple.

A clean workspace allows our devices to be happy, because they are not competing with objects not meant for that space. A clean workspace is dirt and dust free, which allows our computers to remain dirt and dust free. A clean workspace is trash free to keep bits and pieces of debris from getting sucked into a fan/vent of a digital device.

We must work hard to keep our workspaces clean for the sake of our mindscape and digital devices.

Once we have built or bought a PC and learned how to maintain and clean our computer's and the workspaces they occupy. We should be concerned with and learn how to protect our computers and the data stored on them.

CYBER SECURITY

WHAT'S WITH CYBER SECURITY

In general, security is a constant concern of anybody owning property. It is a dance and affair between attackers versus defenders. There is really no such thing as universal or perfect security. It is close to impossible to create a fully locked-down environment. Being free from the threat of an attacker's focus is a fantasy. It does not matter whether we are talking about the digital or physical world. Attackers are opportunist, after all. In cyber security, the reality is no different. We can never really make sure our data is safe. We can only minimize the chance that some hackers or malicious actors get their hands on it.

THIS APPLIES TO OUR DAILY LIFE

Everything we will be talking about, under this big umbrella of cyber security, relates to our daily life. There are so many topics, concepts, and studies to go over in cyber security. It would take years to go through the details. But we are here to build out a toolkit and mindset for all things tech. In our toolkit, we must know how to protect against and be knowledgeable of the dangers involved with being a digital citizen.

WHAT ISN'T HACKING

First, we have got to talk about hackers. Ever seen a movie or TV show with hackers? They will have a million screens typing on a single keyboard, while somehow controlling them all at the same time. They will have an eureka moment, then utter the classic phrase 'ah…I'm in'. While rejoicing they somehow got into a computer network in a mere few moments of time. Well, I got some bad news. That ain't hacking. Hacking is a lot of things. But it is not what we commonly see on the big or small screens. Hacking is not having someone's login information, because they gave you their password years ago. Hacking is not posting a status on your friend's social media account, when they are away from their device. Hacking is not connecting to your neighbor's open Wi-Fi network. Those are more along the lines of finding a dollar on an empty street. Nobody would scream 'thief' if the dollar were picked up. So, what is hacking exactly?

WHAT IS HACKING

Hacking is stealing candy from a baby. Everybody would scream 'thief' and see the action as bad. Hacking is intentional. Hacking is malicious. Hacking can be completely summed up as actively trying to compromise, infiltrate, and target digital devices and networks.

IN THE MIND OF A HACKER

Nothing on the internet is completely secure from a hacker's touch. There is no such thing as perfect security. And the reason is really simple. If a hacker has time, resources, and motivation, they will eventually find a way into the places they do not belong. Hacking is not a fast process. Hacking is not magic.

HACKERS BE HACKING

Hackers are starting to become more sophisticated and use complex means to accomplish their goals. Every day the difficulty of identifying red flags, inconsistencies, and warning signs -- of a cyber attack -- is increasing. Because hackers are hacking.

The Federal Bureau of Investigation, in 2019, recorded 23,775 complaints about business email compromise. Business email compromise is when a hacker has access to email accounts within an organization. This alone accounted for $1.1 billion in losses. In the bigger picture, internet crime and attacks have caused a total of $3.5 billion worth of losses in the United States of America alone. It will be progressively more costly as the years go by because hackers be hacking.

ETHICAL HACKING DOES EXIST

In the Wild West of the Internet, there also exist good bad guys, more formerly known as ethical hackers. Ethical hackers are a necessary line of defense against malicious attackers. They act much like a typical hacker, working to expose system vulnerabilities. However, they do not seek personal gain. Ethical hackers work with many companies and organizations to help boost their defenses. They provide vulnerability assessments and the much-needed mitigation to those vulnerabilities. Ethical hackers be protecting.

KNOWING SOME VULNERABILITIES

In order to protect our digital information, we must get a bit familiar with the ways our devices can be vulnerable. We must get familiar with how attackers think. We must get familiar with some of the methods they use and how they use them. Why is this important? Think about it this way: if you wanted to protect your house from robbers, would you wonder how a robber thinks and what could make you a target? We cannot defend from attackers if we do not know the ways we are vulnerable.

MY FIRST: STUXNET

We have probably all seen the news when some big cyber attack has occurred. Maybe we did not know what it really meant or when it happened. We might not have understood the nuances and details of newscasters and experts explaining it. Nevertheless, we have likely all witnessed a cyber attack affecting millions. And this will be somewhat commonplace moving forward into the future. Heck, I even remember the first major cyber attack of my lifetime. It was Stuxnet.

WHAT IS STUXNET

Stuxnet was a malicious computer software and it was kind of a beast. It was first discovered in 2010. It targeted these things called SCADA systems. SCADA systems are supervisory control and data acquisition systems. Though not overly important, SCADA deals with things related to telecommunications, traffic systems, assembly line machinery, and other similar systems.

Stuxnet was designed to disrupt the SCADA systems in Iran. More precisely, Stuxnet is thought to be responsible for damage done to an Iranian nuclear facility, infecting multiple computers and damaging centrifuges while limiting uranium enrichment. Simply put: Stuxnet. Was. A. Beast.

Stuxnet showed the world cyber security and the threat cyber attacks present cannot be ignored. It was the world's first digital weapon and the first cyber attack I personally remember. Stuxnet's significance cannot be understated.

WHAT THIS MEANS TO US?

Cyber attacks existed before Stuxnet and have definitely existed after. But the world had never seen such an attack of this type and scale. It made nations and companies take notice. But it also meant something to the normal end user, for us.

The whole attack started with a universal serial bus drive or USB, to keep it simple. Most of us are familiar with USBs or something similar. We use them to store data, work files, or even cryptocurrency. But the convenience of easy and simple data transfer and storage has a dark side. We might not see USBs as a vulnerability, however any digital device can hold malicious software ready for attack and we could be unaware of it. The most eye-opening part is that we often do not even realize common items and their usage can be the attacker's way into sensitive areas of our life.

We do not need to be a nation, state, or major company to be affected by malware. We do not need to have millions of dollars to become a target. We do not even need to be connected to the internet to feel the effects of a cyber attack. Everybody has something of value to the right attacker, especially if they are looking into the organizations and companies we work for and frequent.

THE DETAILS ON MALWARE

Malware is any software written with malicious intent. Malware seeks to damage, corrupt, or degrade. Malware can be made to target any type of digital device. So, it is important to know which ones can affect us in our daily lives like ransomware, spyware, rootkits, and many more. Let us learn about a few and their differences.

RANSOMWARE

Around 2016, there was a surge of malware targeting hospitals. This malware is called ransomware. Ransomware, like the name implies, seeks to hold a computer's resources and data hostage for a ransom. If the ransom is not paid, then the information is released or destroyed.

Ransomware is a nasty form of malware because the main targets are hospitals and their sensitive data related to healthcare. I personally cannot think of a worse way to harm people without inducing physical threat.

Imagine you are in a doctor's office and they cannot access your medical records or que you up for a much-needed surgery. That could be devastating and deadly. This can all be caused by the ability to lock users out of their computers. That is ransomware.

SPYWARE

Sticking with the naming tradition and convention, spyware is simply software designed to spy. This type of software secretly collects information from our digital devices. It keeps track of every action taken on our devices, then it sends that data back to a remote computer/hacker. Spyware will especially watch network traffic and web browsing habits.

BRUTE FORCE

If you have email, banking, or social media accounts then you probably know about the crazy requirements for passwords, frustrating as they can be. Those requirements are necessary. They help protect us from brute force attacks.

Brute force attacks or brute forcing is a hacker trying to guess your passwords. It might be better to call it educated guessing, because a hacker can use a bit of basic logic for figuring out someone's password. The logic used revolves around human habit. People typically choose passwords related to them. For example, a family member's name, job, hobbies, or other similar concepts.

There are even a few different techniques to using a brute force attack, but we are mainly concerned about dictionary brute forcing.

DICTIONARY ATTACK

A dictionary attack takes a username or email and tries it against a list of commonly used passwords. This is done for the sake of gaining access to our computers, networks, and private information. Not to worry though, having a strong password is the simplest way to defend against this type of attack.

PASSWORDS SUCK

Making passwords can be a drag, to the point where people even outsource the creation and storing of passwords to other programs. Some of people even use the same password for all of their accounts. I personally recommend against both of those methods in the name of security. Here is why.

Allowing another service to create passwords and store them for you gives a third party too much access to your personal and private information. Simply put, we could ultimately lose control and privacy. Also, what if that program or service is hacked?

If we use the same password for all of our accounts, then this makes it easy for a successful hacker to hack all accounts using the same combination of email/username and password. Simply put, an attacker with access to one account has access to all accounts.

Passwords suck, especially for forgetful people like me, but they are a necessary tool – which is why we should all have strong passwords. So, what does a strong password look like?

A PASSWORD SHOULD LOOK LIKE

Here is what a strong password should look like:

- Contain 8 or more characters—more is better
- Mixture of UPPERCASE and lowercase
- Mixture of letter and numb3rs
- Include at least one special character, e.g., ! @ # ?]

Here is how to keep it secure:

- Change or rotate passwords regularly
- Do not use common words
- Do not use repeated characters, e.g., AAAAA or 12345
- Do not use personal information e.g., names of family, pets, friends, social security numbers, birthdays, etc.
- Do not write a password and store it near a computer

PHISHING

One of the easiest ways for hackers to get into our networks and personal information is phishing. Phishing is when a hacker sends email or private messages attempting to get sensitive information. Examples of sensitive information are usernames, passwords, company details, or bank account details. This information can be anything that helps the hacker accomplish their goal. Whatever that goal might be.

Phishing is the most common tool in a hacker's tool belt because it preys on human psychology. A good phishing attack appears to come from a trusted source. A good phishing attack uses public information about us to build rapport. Those types of techniques are called social engineering.

Phishing attacks also might involve an attachment via email or private message which loads malware onto your computer when clicked or opened. So be cautious and suspicious of unexpected emails with offers that seem too good to be true or that prey on any potential life insecurities. Phishing could be a link to an illegitimate website that looks similar to a trusted link. For example, you thought the link said Netflix.com but really it said NTflix.com. When clicked the fake website can force the download of malware and other types of software. It might start asking for personal or sensitive information. It could redirect you to another website a place you did not intend to go, so be diligent in spotting phishing attacks.

DISTRIBUTED DENIAL OF SERVICE

Distributed Denial of Service or DDoS seeks to overload the capacity of a network, computer, or other network connected systems. This is done so the system cannot respond to service request received. The best way to describe this attack is – it is like a four-year-old asking 200 questions in five minutes. There is no reasonable way to respond to all of those questions, so we just shut down or respond with short answers without answering the questions or fulfilling the request. DDoS attacks cause the same effect on computer systems.

BOTNETS

The last type of cyberattack we will talk about is a botnet attack. Botnets are a network of devices infected with malware. They are often used for DDoS attacks by taking hundreds to thousands of computers forming a collective to perform a single unified attack. Typically, they target all the same systems at the same time. Botnet attacks are close to untraceable because they comprise so many different computers. Computers in this style of attack can be from multiple locations. Since there are numerous attack computers tracing each individual infected computer is nearly impossible.

IT'S NOT ALL BAD

Everything is not doom and gloom. Yes, there are tons of attacks hackers use to compromise digital devices and networks. But cyber security professionals work just as hard to prevent and mitigate hacker's attacks. Cyber security professionals have developed numerous defensive techniques and tools. The standard for creating a strong password is one such tool. Let us explore more of them.

TWO FACTOR AUTHENTICATION

It is easy to look at two factor authentication as an unnecessary step for logging into our apps or websites. But it is an important addition to our cyber security safety net. It is an added layer to our already strong password that we learned to make earlier. How does it work?

Two factor authentication is a process that requires two pieces of data proving we are who we say we are. This gives protection from hackers if they somehow acquire our passwords. It is that simple. But two factor authentication is not the highest form of verification around.

MULTI FACTOR AUTHENTICATION

Above two factor authentication, there exists three factor authentication. We call this multi factor authentication, and two factor authentication is included.

For proving who we are, there are three commonly accepted factors of authentication. They are:

- Something we know
    - Passwords or pin numbers
- Something we have
    - ID cards or physical keys
- Something we are
    - Face, fingerprints, or retinas

When using two factors like a password and ID card is two factor authentication. Using all three is three factor authentication.

PAUSE AND TAKE TIME

It is easy to get excited when installing a new app or making an account for a new website. Especially if it is something fun and useful to our lives. But take your time and read everything when installing apps and making accounts. Read the terms of service, privacy policy, and for e-stores or online purchases, the refund policy. Reading these helps us understand what exactly we opt into when using online services. They allow us to know which data is required for the service to work perfectly. But most importantly, this information lets us see if we can opt out of giving unnecessary data about ourselves. So, pause and take time to read.

KEEP THOSE DEVICE UPDATED

Computers and other electronic devices receive frequent software updates. This might seem annoying, because updates might offer a single feature no one will really use or a cosmetic change we might think was needless or ugly. Yet, updates typically do more than what we see on the surface level.

Software updates help our devices defend against attackers. Companies are always looking to protect their products and the users that enjoy them, because if a device is risky or vulnerable, consumers are less likely to purchase it again.

Patches for vulnerabilities are typically contained in software updates, so it is vital to our cyber security to keep our devices updated. Yes, updates can take a small amount of time to download and install. But being protected is much more important than a few lost minutes.

IS THAT ME

Computers can have errors and these errors are normally linked with some numeric code to identify them. But there exists an error so heinous most people dare not speak its name. It is disgracefully known as user error. User error is an error made by human input, i.e., a person clicking, typing, or connecting the wrong thing. Sometimes user error can be the hardest to notice when there is no outside perspective. It is the most common error when dealing with anything tech related because we might not realize we are doing something wrong. But it is also one of the easiest errors to fix and to be aware of. Simply ask yourself this question: did I make a mistake?

WHAT TO DO IF HACKED

A hacker could steal credit information, create a malware that simply steals a portion of a computer system's resources to accomplish a bigger goal, or even access your webcam. If you are hacked, it is a terrible and frustrating experience. But it is not the end of the world.

What to do when hacked:

1. Share that you have been hacked with family and friends to let them know certain emails or messages from you might not be authentic.
2. It might be worth doing a credit freeze and locking down all assets.
3. Request new debit, credit, and savings cards from your bank(s).
4. Factory reset all network-attached devices, i.e.: computers, phones, router.
5. Change all passwords to stronger passwords. Turn on two factor authentication where possible.
6. Deauthorize/deactivate third party apps connected to your social media, email, and other similar accounts.
7. Monitor all your accounts closely over the next couple of months.

GOOD HABITS TO HAVE

Knowing what to do before getting hacked is equally as important as knowing what to do after. This list will not prevent hacking, but it will help reduce the risk a bit.

1. Install antivirus and anti-spyware software.
2. Be cautious about clicking on pop-up boxes.
3. Be cautious of unfamiliar and untrusted emails, especially if they have an attachment.
4. Keep all devices software up to date.
5. Install a firewall.
6. Lockdown your router by changing the default password to a strong password. Use good password habits.
7. Change the name of your router from the default name.
8. For your router security, choose the WPA3 if available, but WPA2 or WPA will do as well (more on WPA later).
9. Do not trust free Wi-Fi, a hacker can see and monitor network traffic for sensitive information. Remember, if it is free for you to use then it is free for a hacker to use.

10. Check bank statements for anomalies.
11. Make regular backups of your important files.
12. Never give personal information to untrusted sites or persons.

ANTIVIRUS, FIREWALLS, ENCRYPTION

Before ending this section on cyber security, there are three important tools I want to discuss. They are antivirus, firewalls, and encryption. Each one of these tools serves a purpose in securing our information and digital devices.

Antivirus is a type of software designed to find, prevent, and remove malicious computer software. Majority of antivirus works on something called signatures. A signature is essentially the fingerprint of malware and is an easy way for antivirus to find, locate, and ultimately eliminate a virus. Most antivirus passively scans and removes malware but may require some limited interaction from the computer user. Antivirus is vital for computer security and protecting personal information on digital devices. Equip every device you have with antivirus, where and whenever possible.

Firewalls can be software or a hardware device. Firewalls control which network traffic is allowed and denied. They do not identify or remove anything. They simply block or accept. For example, it is possible with a firewall to block all network traffic coming from a specific IP address.

A router is a networking device which routes traffic between computers networks. Routers help identify what IP address is connected to what computer. Routers also use firewalls to block unwanted internet traffic. If you have internet service in your house, then you probably have a router. Most routers have a default firewall. It is decent at blocking most unwanted traffic.

Earlier, I mentioned something called WPA. WPA stands for Wi-Fi Protected Access. Every router since the early 2000s has some version of this authentication protocol. As the name implies, WPA was created to help protect access to Wi-Fi networks. It does this through encryption.

There are many different types of encryption. We will go over one later, VPN encryption. For now, let us look at encryption in the broader sense. Encryption is basically passing information from point to another using an agreed upon code or key. Most information in the modern day is encrypted and encrypted data is called ciphertext. Unencrypted data is referred to as plaintext. Cryptography is the study of hiding information from outside observers, i.e. encrypting and decrypting.

I highlighted the key concepts of antivirus, firewalls, and encryption, because they help us stay more secure in a complex world of technology, information, and hackers. Each one plays a role in our daily lives. In order to stay safer while enjoying the internet, it is best to invest time in knowing the tools which enable that.

COVID-19

When COVID-19 hit in 2020, tons of people were forced to work from home. This led to a massive amount of people having to learn and adapt to new technology, working conditions and the stressors they bring. No technological tools became more useful to the collective of new remote workers than VPNs and video conferencing. Let us take a brief look at VPNs: what they are, how they work and why we need them.

VPNS

VPN stands for Virtual Private Network. As the name implies, a VPN provides a virtualized network to access the internet safely and securely, helping us keep our private information private.

A VPN works by directing all of our internet traffic through a secure virtual tunnel. Imagine normal internet traffic like driving in the rain trying to get from point A to point B. The cars are going to get wet no matter how fast our windshield wipers move. Now imagine driving in the rain through a tunnel. We would not even notice it was raining. That is the power of a VPN. It keeps our data from prying eyes. The same way tunnels keep our cars dry in bad weather.

VPNs are used to secure data. Most companies use them to protect their data for worker's working offsite or at home. But VPNs are useful to anyone not using a trusted network. A trusted network simply means a network that is password protected, private (i.e., home, personal), and free from unknown users. The only network most people can consider trusted is our home networks. We know it is password protected. We know it is private. We know it is free from unknown users. Therefore, it is trusted. But when not using a trusted network always and I mean always use a VPN.

The usefulness of a virtual private network does not stop at securing our private data. A VPN can also act as a proxy server. A proxy server makes network traffic appear to be coming from a different location. Meaning if a Netflix user from the Unites States of America (US) is in Japan and wants to access their favorite shows they get back home. Then they would need to find a proxy server with a US based Internet Protocol (IP) address.

WHERE TO GET A VPN

There are lots of places selling a Virtual Private Network. And no discredit to any of them. They provide a useful and much needed service to people not technically minded or who like to keep things plug and play. But a VPN is one of few things I take a hard stance on. It is always best to make our own VPNs. Using our own home networks. Because they are the most trusted networks to use. We can also verify our data is not being sold to a third-party vendors.

Making a VPN is not necessarily difficult, and it is actually inexpensive. But it does take following some instructions nearly word for word. There are lots of great examples of how to set up a VPN on a home network. I will not go into the details, because it is a bit out of scope of this book. I can say this. Anyone making a VPN at home will need two things:

1. A device that will never disconnect from your home network. It could be an old laptop, desktop, or Raspberry Pi.

2. A good deal of patience and willingness to learn and troubleshoot

Trust me the results are worth it. Knowing your data and private information are secure feels amazing.

If you are left wondering, what to type in a search engine to learn how to make a personal VPN. Here is some help. Search: **vpn raspberry pi** or **vpn old laptop**. Those will get you going in the right direction.

OPTIMIZE SEARCHES

WHAT MADE GOOGLE DIFFERENT

Google is a unique beast. It rose to and claimed success in a saturated market with a lot of big-name competitors. Originally, Google's main goal was to design a simple search engine to organize information and make it useful and easily accessible. Other search engines of the 80s to mid-90s struggled with keeping things simple. They would often display the weather, some news, and sprinkle in a few advertisements all on a single web page.

A web page is a page on a website. A website is an organized group of web pages. It is possible to have a website with just one web page. To view any website requires a web browser. A web browser is a software that allows access to the World Wide Web. The World Wide Web is simply the internet.

Google competitors web pages were bloated with a ton of content, which meant every search or web page loading was slower. This created an unpleasant user experience. But Google was different.

Google understood the importance of keeping it simple. Too much content on one web page can be distracting for users. Google did something most of its competitors would not do; it simplified.

BETTER INTERNET SEARCHES

Since we talked a bit about Google earlier, let us learn how to get the most out of our internet searches. These tips are primarily for Google, but some might work for other search engines. Here is an introduction to some of the ways in which we can simplify and optimize searches.

KEEP SEARCHES SIMPLE

Most search engines know how to search for tons of different information. So that means we do not have to be specific. We just have to feed search engines simple information and those fancy algorithms start going to work.

What are some good examples?

1. Need your fix for cute kittens in teacups? Then simply search for: **kitten in teacup**
2. Looking for your local cat cafe? Then you can easily search using: **cat cafe nearby**
    - A lot of search engines, especially those with a GPS web mapping application, will try to grab your relative location and display search results nearest to you.

SEARCHING SITES

There are these things called operators. They help us when performing searches on Google. Essentially, super charging our web searches by allowing us to do specific things. There are quite a few of them. Let us start by talking about the operator **site:** .

The **site:** operator gives the power to limit search results to only the website in mind.

How to use it? It is really simple. Say you get the craving for some cat-shaped cookies, you do not have a cat-shaped cookie cutter, and you only want to buy from Amazon. Type in the search bar: **cat shaped cookie cutter site:amazon.com** and hit that magical enter button. And boom! All of the results will be from Amazon and cat cookie cutter related. The syntax: **<thing you want to search> site:<website you want results from>**. Syntax is the structure in which we order words or characters to accomplish our desired result.

LOOK FOR SPECIFIC FILES

One of the more underknown operators is **filetype:**. It is also one of the more useful operators to know. **Filetype:** gives the ability to search for files and/or types of files.

The **filetype:** operator is especially powerful in the corporate setting. Imagine if you have been tasked with creating a PowerPoint, but you remember a similar one your company made years ago. With the **filetpye:** operator you can find it in no time. All you have to do is use this syntax: **<what you want to search> filetype:<file type you are looking for>**.

USE THAT WILDCARD Y'ALL

There is this thing called a wildcard. It is not necessarily an operator. In a Google search, we can use an asterisk as a wildcard. A wildcard is a really powerful tool. Wildcards act as a placeholder and search engines will try to fill in the placeholder. If you are ever trying to think of a song or a quote, then the asterisk wildcard is here to save the day. This is a great way to fill in those blank spots. Simply use this syntax: **<words you remember> * <more words you remember>**. Example: **what the * is cyber**.

ONLY SEARCH FOR WORDS YOU WANT

When you are not at a loss for the word or words you need to search, there is also an operator(s) for that. It is quotation marks. These nifty little dillies **""**. Wrap whatever you want to search in a set of quotation marks and the results will only contain those words. The syntax is very simple: "<something you want to search>" Example: **"I like kittens"**.

Searching **I like kittens** without the quotation marks operator produces around 143,000,000 results. While **"I like kittens"** brings it down to nearly 450, 000 results. 450,000 is still a lot. But at least we know everything will be related to only what we want to search.

LOOK FOR TWO AT ONCE

OR is probably the simplest operator to understand. It allows us to search for one word/phrase while also searching for another. Really useful if you ever want to search 'are cats cool' and 'dog like cats' at the same time. It is like looking left and right at the same time.

How to use it? The syntax requires using two sets of quotation marks (**""**) and placing an OR in between the items we want to search. It looks something like this: **"are cats cool" OR "dog like cats"**. This will bring up results related to either 'are cats cool' or 'dog like cats'. The syntax: **"<thing you want to search>" OR "<another thing you want to search >"**.

SOME WORDS GET THROWN AWAY

Most search engines actually throw away words or remove them from your search completely. These are called stop words. Removing stop words allows for fast searching, which means we get those sweet cat pictures that much faster.

What are some stop words we might be quietly pondering to ourselves? This is not an inclusive and all-encompassing list. But some stop words are:

- A
- AN
- THE
- IN
- OF

Why is this important? Sometimes we can get some funky looking results and stop words might be the reason. We might need to employ other methods and tips for searching for the content we want. On the other side of the coin, stop words are often removed because they do not affect the search result. So, we do not always have to stress about describing and including all the 'the's and 'of's. Because 'the greatest cat of the world' and 'the greatest cat world' return near identical results.

DIGITAL MINIMALISM

ALWAYS KEEP IT SIMPLE

Digital life can get complex and convoluted. We might not always notice the little changes, because technology is so ingrained in our lives.

It is easy to lose self in the flow and be swept away. It is easy to feel shocked by everything going on in the world of technology and the world in general. It can be difficult to take a step back and remember who we are without our devices, especially if you are following your natural sense of curiosity. But not to worry, there is a solution for the ever-growing complexity of the digital world. It is called digital minimalism.

WHAT IS DIGITAL MINIMALISM

Digital minimalism is not an overly complicated concept; especially when we break up the two words. Digital represents all things cyber and computer related. Minimalism represents living with less but getting more out of what we already have. Combine them together and we have digital minimalism, which can be summed up as the digital pursuit of less. It is all about getting more out of your technology without increasing your digital footprint.

IS DIGITAL MINIMALISM IMPORTANT

Digital minimalism is important. It is almost vital. For the average world citizen, digital technologies will touch some part of life. Our lives are affected by technology sometimes directly and sometimes indirectly.

Sometimes we do not even notice how much digital technology plays a part of our life. For instance, if you live in a modern city, you might own 10 or more smart devices. And it is okay not knowing exactly how many you have. Smart devices are meant to seamlessly integrate into our lives. Providing functionality while hiding in plain sight.

A great example of smart technology hiding in plain sight is an internet-connected washer or dryer, coffee maker, toothbrush, or even bed. These are not items commonly associated with the internet or cyber, but they can be. It can be interesting at times to think about. We do not always know the level our lives are shaped by the environment and technology around us.

With our lives becoming hyper-connected and technology-centric, we must find a way to remain human, centered, and uncontrolled by technology. Digital minimalism allows for this and getting more out of our cyber experience.

WHY AM I A DIGITAL MINIMALIST

I am an accidental minimalist but an intentional digital minimalist. I stumbled across minimalism when a friend mentioned it to me. She randomly compared my lifestyle to that of a minimalist, so I started a frantic search for all the information I could find on the subject. I probably looked like a madman trying to consume all the knowledge, but she was right. On the surface, I looked and lived like a minimalist. For example, I prided myself on having only what is useful, keeping what makes me happy, and attempted to be efficient in all that I do. But I did not, or rather I could not, find minimalism in one part of my life, the digital part.

I was not using my technology intentionally. I was not researching my digital purchases. I was not simplifying my digital life. It was complicated. My desktop and phone had tons of unused apps. I had tons of photos saved in my phones, hard drives and in the cloud. I was just hopelessly clinging to old photos, unused phones, and other digital devices. I needed to change. I needed something different in my life. I needed digital minimalism.

IS DIGITAL MINIMALISM FOR YOU

I honestly cannot tell anybody if digital minimalism is for them or not. But I often tell people that learning different ways to view technology and how we interact with it can be helpful. Digital minimalism can help when deciding whether or not to get that new gadget you have been eyeing. Digital minimalism can help keep you in touch with the younger generations and support a better future for them. It can help you get more out of the devices you already own. A little knowledge about something new can never hurt.

A NEED FOR DIGITAL MINIMALISM

Most of us might be involuntary digital hoarders. This just seems to come with access to a cell phone or computer. Almost like a force of nature making some decree: if there is technology, then the people shall hoard.

Majority of the time this form of hoarding goes unnoticed. We often cannot touch, taste, or smell the items cluttering up our digital spaces. The need for digital minimalism lives next to its crafty neighbors, the two types of digital hoarding.

TWO TYPES OF DIGITAL HOARDING

We live in two different worlds. In these worlds, there exist two types of digital hoarding. One is obvious and the other is sneaky. In the physical world, we might be afraid to throw away our old cell phones, cables, and remote controls we never use. In the digital world, we may take 100 pictures and keep them all or download apps only to use them once and struggle to delete them for months or years. We are probably all struggling with a bit of digital hoarding, whether it is in the digital or physical world. That is okay, because we are rarely taught how or why we should tidy up our digital spaces on our devices.

FIRST TYPE OF DIGITAL HOARDING

The first type of digital hoarding you might be familiar with. There have been tons of books written and many television shows produced on the subject, as well as podcasts discussing fascinating aspects of it. The first type of digital hoarding is simply regular hoarding of digital devices.

We have become obsessed with getting the latest gadgets. We rush to stores or fill our online shopping carts with new tech, often forgetting that we already have devices for our day-to-day needs.

We are then left with all these old devices, we inexplicably cannot rid ourselves of, and our new devices that are soon to be replaced when the next new thing catches our eye.

Are you a physical digital hoarder?

SECOND TYPE OF DIGITAL HOARDING

The second type of digital hoarding is almost invisible and less likely to be understood. We are not trained or taught to notice it, because we do not see the reality and effects of this type of hoarding. We do not see the massive data centers holding all the backups of cherished memories and files we refuse to delete. We do not find problematic the hundreds of selfies locked away on our phones we rarely revisit. We have become oddly okay with having 1,000 open tabs on our web browser, even when we cannot pin down which one is playing the audio we hear. We are failing to realize that this is a form of hoarding, because it is near invisible in our daily lives – involuntary digital hoarding.

MY FOUR GOLDEN GUIDES

When I discovered my minimalist ways in 2016, I also started my digital minimalism journey. Like any other journey, I gave myself some tools to use if I ever felt the journey became too tough or I started to stray from the path. These are helpers and not laws. They are meant to assist and not hinder. I present my Four Golden Guides.

LIFESTYLE AND DIGITAL MINIMALISM

Guide Number 1: Know our lifestyle and what fits into it.

No one can accurately describe us, our lives, and our needs better than ourselves. This means we know what is necessary and unnecessary in our life. We know when a purchase is bad for us and when one is good. We know when something adds value to our life and when something subtracts or distracts from our life. This is why it is important to know what fits into our lifestyle, so we are free from the burden of the unnecessary and unessential.

RESEARCH AND DIGITAL MINIMALISM

Guide Number 2: Research what we buy, download or plan to use, so it syncs with our life.

Knowing our lifestyle and what fits into it is important. But if we never research what we buy, download or plan to use, then how can we truly understand how these items fit into our life? At best, we could become uninformed consumers without a bit of research. There is a common perception that people consume by taking the advertiser's word at face value and not independently finding out anything more information. The following question kind of highlights the concept. What is the difference between an expensive versus cheap HDMI cable? The answer is nothing and a lot of times that is the case. Certain digital technologies have gotten to the point where cheap and expensive items like smart watches, TVs, and phones, for example. Do not really suffer quality wise from the difference in price points. But this might not be known without some investigation.

In order to get more out of what we have, we must investigate our needs and the things promising to fulfill them. There is nothing worse than buying a 50-inch television only to realize when you get it home, unpackage it and set it up that we needed a 60-inch TV with a soundbar. We must become better informed consumers of digital devices and virtual shoppers.

Being an informed consumer of technology centers around one concept research. Researching entails looking deeper than information given by advertisers. Using a search engine like Google or Bing, we can compare function, price, and reviews in a fraction of time it would take us to call or visit multiple physical stores.

We must also research utility to our life. We need to research the benefits of a potential purchase, like whether we even needed a vacuum cleaner, for example given the particulars of our lives.

CONNECTION AND DIGITAL MINIMALISM

Guide Number 3: Physical connection is greater than virtual.

We are more connected than ever before. It may be a cliche statement, but it is true. In less than a couple of seconds we can visit faraway places and make friends. We can even turn lights on and off in a room we have never been to before, located on the other side of the world. Distance has become a trivial thing, to the point where many of our social interactions start in the virtual world first. Whether it is dating, gaming, job hunting or buying a house. The process will likely start with a digital interaction first. This is not a bad thing, this is a somewhat natural progress in the world of technology and be more connected than ever before. Which is why we have to remember that physical connection is greater than virtual.

By this I do not mean abandon all technology and become a wild person in a forest somewhere. I simply mean to value an in-person connection more highly than a digital one.

Humans are social creatures and we cultivated skills to engage, interact and be stimulated via face-to-face in person interactions. We have spent decades, centuries and millennia mastering the dance of conversation and in person communication and the awkwardness it often provides. Because even when we are bad at communicating in person, we are still fulfilling a lot of the necessary components of a successful interaction. Like the nuances of reading someone's facial expression or their body language, these are things that are often missed or lacking in digital communication.

Though, digital communication can be less personal. When we are deeply familiar with the persons we communicate with. Digital communication can be amazing for maintaining social connections.

BREAKS AND DIGITAL MINIMALISM

Guide Number 4: Take breaks from technology.

It is important to take breaks from technology, as breaks allow for time to refocus on life and realize what is important and essential. It is easy to get sidetracked by the constant flow of daily information streamed digitally. Life can be so busy, funny, distracting, and somber. So, it is vital to take breaks from technology and social media, because they can often amplify the effects of what life delivers. Some breaks might be purposeful and intentional, while some breaks might be abrupt.

I have taken many intentional breaks. For example:

I like to create music. When I go into music creation mode, I isolate myself from technology revolving around music like music streaming apps. I do not listen to the radio. I do not stream any songs. I do not answer or open any music-related emails. I like to establish a vacuum. If it is not involved in the music-making process, then it does not get in.

I have also had my good share of unintentional breaks. For example:

When I got into a car accident and could not engage properly with technology or type a coherent sentence, I took a pause.

Life just happens, whether it is something in the physical or virtual world. Life can give us so many reasons to take a break from technology or slow down our use of it. And that is okay.

KEEP WHAT MAKES US HAPPY

In our pursuit of digital minimalism, it is often the aspirational goal to have no or very few electronic devices. But it is important to keep what makes us happy. We could become miserable trying to whittle down our electronic life to the necessities. Our mission in digital minimalism is happiness and living with what fits our lives, while making sure not to thoughtlessly cram items into our life without purpose. Always keep what makes you happy. Get rid of what does not. Try a bit of digital decluttering.

START DIGITAL DECLUTTERING

Digital decluttering is spring cleaning for our digital devices. Much like real life clutter, we must unclog our unseen storage of data and free up space. Decluttering is a bit of a journey, but the end result of an organized digital life is worth it.

DECLUTTERING AND BACKUPS

The first step on any digital decluttering journey is creating a backup. Backups provide redundancy for our digital files the essence of our digital life. Backups allow us a baseline for security in case we need to revert to a safe state after being hacked or infected with a virus. Backups allows for peace of mind in case we delete an important on our main devices. Backups allows for stress free decluttering.

There are two ways to create backups -- external hard drive or virtual cloud. External hard drives deliver an excellent way to store data locally. Meaning your backups are a bit more secure, because only you will have access to them. Virtual clouds are a great way of storing data that is accessible from anywhere. Virtual clouds are a network of computers called servers. Servers store and serve data to authorized users. Both ways of creating backups have positives and negatives.

External hard drives are fairly cheap and have massive storage capacity. One negative is lack of redundancy. Meaning the data is only backed up to one storage device. Another negative is lack of remote access. Meaning in order to retrieve any information from an external hard drive it has to be physically on your person.

Virtual cloud storage, typically, has built in redundancy and remote accessibility from anywhere. Meaning data is stored on more than one server or storage device. One negative is cost. Cloud storage is expensive if you want to store a lot of data.

Deciding if you should use external hard drives or cloud storage can be tough. I would recommend a hybrid approach. Using external hard drives for the most important documents like taxes, contracts, etc.; while using cloud storage for pictures, homework, grocery list, etc.

How often you backup your files is subjective. I, personally, back mine up once or twice a year when I am in super declutter mode. This rate of backing up helps me reduce cluttering my external hard drive with unnecessary backups.

DECLUTTERING AND ROUTINES

Routines keep us in the tidying groove. Like cleaning a house. It is good to keep a schedule for decluttering our digital spaces. Good routines build habits. Set a to-do list to review and get rid of pictures, files, and even devices that do not bring happiness or have use. I personally do this once a week. Decluttering monthly is fine as well.

DECLUTTERING AND NAMING

Something which aides decluttering and staying tidy in the digital space is having a naming convention. A naming convention is a rule for how to name files, folders, etc. A naming convention allows a file's name to provide useful information on its contents.

An example of a naming convention is a file named 2018Oct_Grandma_Atlanta. This naming convention lets me know three things: when, who/what and where. The underscores also allow for separation of categories. For personal use, this is an adequate naming convention. Though, it might not work for every individual. It is a good template.

Naming conventions are useful for helping to understand the contents of a file or folder without opening them up. This will definitely make life easier when looking for documents and files. Please take the time to purposefully name each file while following a naming convention if you do not already.

DECLUTTERING AND DESKTOPS

Is your virtual desktop litter with icons, files, and apps? When was the last time you cleaned your desktop? It might be time to clean up.

In order to keep tidy, it best not to save or place all of your files on your desktop. In the spirit of decluttering, create folders to organize your desktop into sections and themes. You could create folders for finance, games, photos, etc. Whatever you need to stay organized. Then enjoy a visually organized and noticeably cleaner desktop.

Emptying your recycling bin on your desktop can also be beneficial. It frees up space and provides instant gratification of decluttering by deleting old forgotten files.

DECLUTTERING AND DOWNLOADS

Often overlooked is the downloads folder. Emptying your download folder creates tons of room on computers. Because years can go by before we ever consider deleting anything within the downloads folder.

DECLUTTERING AND SORT BY

Computers allow for a number of different ways to sort information. You can sort by date, file size, file type, and name. In the pursuit to declutter and free up digital space, there are two vital sort options, sort by size and sort date. Sorting by size allows for deleting file taking up the most space. While sorting by date allows for getting rid of those pesky old files.

To not feel overwhelmed with the herculean task of decluttering, these two options allow you to focus on what is important. Freeing up space and getting rid of older unnecessary files while keeping what is essential to our digital lives.

DECLUTTERING AND EMAILS

One exemption to decluttering weekly or monthly is email. Emails accounts are bombarded daily with spam, subscriptions, updates, and even phishing attacks. Few emails hold important information. Yes, spam filters exist, but they are not perfect.

You must sort through and delete email daily, because sorting through thousands of emails is not for the faint of heart. However, if you are attempting to sort through thousands of emails, then it is not the end of the world.

Getting your email under control takes time but is worth the time taken. First, go through your email and delete all the unnecessary information. Focus on sale offers, subscriptions, and emails from random people and websites. Most email services provide a search function for email accounts. This empowers you to find emails in a precise manner. For instance, if you are getting emails from TidyCyber.com, but you do not want them cluttering up your inbox. You easily type TidyCyber.com into the search function. It will group allow related emails. Then you can delete all emails of a similar type. Know you can also sort by date and delete the older emails first.

Next, you want to attempt stopping the unnecessary emails at their source. This requires unsubscribing to email lists or deleting accounts on the website/service sending emails. Most email subscriptions have an unsubscribe section at the bottom of the email.

To keep your emails under control, I strongly recommend performing a daily audit. Asking the simple question: is this an important email?

DECLUTTING AND EMAIL ACCOUNTS

Email accounts are an unsung hero of organization and security for our personal lives. Having multiple email accounts can make it difficult for hackers to access our sensitive information because it is compartmentalized. This can be taken further by giving our emails purpose. Each email should serve a function. Have an email for spam, bills, applying for jobs and whatever else. If a hacker gained access to one account, the odds of them getting anything useful is reduced. But it also allows for easy simple organization. What are the bills for the month? Boom...look at the billing email account. Organization and another level of security through emails.

HEALTH AND TECHNOLOGY

TECHNOLOGY-FREE ZONES

With technology becoming more involved and integrated in our daily lives, a new question should be asked toward our societies: should we have technology-free zones?

A technology-free zone (TFZ), as the name implies, is an area free from technology -- free from the reach of social media, cameras and electromagnetic fields generated by wireless signals. These zones might become a new necessity for us, much like how public parks encourage recreation. TFZs could encourage in person socializing, reduce our dependence on tech and boost mental health. It is a concept worth exploring and may become a reality in the future.

OUR BODIES AND TECH

One argument for technology-free zones is the negative physical reaction technology can have on our bodies, like electromagnetic hypersensitivity or EHS. EHS, like the name suggests, is a sensitivity to electromagnetic waves. Some of the most common symptoms are redness of skin, tingling, dizziness, or nausea.

As I understand it, it is believed doctors should not treat EHS as a medical diagnosis, because the symptoms are nonspecific, often differing from person to person. This is not to discredit or invalidate people experiencing symptoms, but it raises two important questions:

- What effect does technology really have on our bodies, short and long term?
- How do we prevent and mitigate those effects and still progress forward with tech?

The answers to the question are not clear yet. We are still actively trying to answer them as a society. The answers might not come quick or without struggle, but we will eventually acquire them. One thing is definitely true, progress waits for no one.

MENTAL HEALTH AND TECH

It is no secret that technology affects our lives externally. It also affects our lives internally. Technology has started to shape our mind and brings out some of the ugly sides of mental wellness. These effects can be long lasting and damaging but are often avoidable.

MENTAL HEALTH AND SOCIAL MEDIA

Social media provides a connection to the world without physical interaction. Social media allows for our art, thoughts, and opinions to be shared instantly. It gives us a chance to become a star. Social media helps people find like-minded individuals. Social media provides an avenue for real world social change and activism. However, social media has a dark side. For every positive thing that social media offers, it also takes something away.

Social media platforms allow for an unprecedented level of human interaction, but there is a price paid via mental health. Social media usage can help mental health issues to appear in perfectly healthy people, often manifesting as depression or anxiety.

Using social media does not always lead to depression or anxiety. Though, there are some correlations. Social media usage can cause self-image and self-esteem issues because we are able to compare other people's lives with our own thus creating feelings of inadequacy. It can prompt a fear of missing out, or FOMO, because social media is fluid and always updating, making us think we could miss the next big thing or trend. This could cause us to constantly feel the need to be attached to our devices, and when we are not attached, feelings of anxiety could arise. Social media can make us feel isolated. It can even make us vain and self-centered because we become used to sharing our personal thoughts and being rewarded via likes and shares. This is the dark side of social media.

This is not a warning to abandoned social media. Social media is important; it keeps us connected to the people we love and the causes we believe in. This is advice to utilize a mindful approach when incorporating social media into your daily social portfolio.

BREAKS FROM TECH ARE IMPORTANT

To reemphasize, humans are social creatures. Social media can never replace the stress-reducing and mood-boosting effects of good old face-to-face interaction. Do not be afraid to disconnect from tech and reconnect to the physical world.

PHANTOM VIBRATIONS

There was a time. Long, long, long, long, long ago...when texting was new. Okay, maybe not that long ago, but still pretty distant from memory for most. Our bodies started to adopt a weird new habit.

Remember phantom vibration syndrome? If not, that is okay. Here is a refresher. Phantom vibration syndrome is a strange phenomenon. This phenomenon would cause people to imagine phone vibrations and sounds. People were hearing and feeling things that were not happening. The stranger part is that most cell phone users have experienced this at some point. Most importantly, this moment in behavior and physical response to tech is us learning a new behavior. Sometimes our bodies and minds do not always work well together with odd new behaviors that technology brings into our lives. So sometimes we just have to embrace the weird and new.

TIME TO PUT DOWN THE PHONE

When your body starts imagining vibrations and sounds, that might seem like the time to put down the phone and step away from technology. However, those responses are generally normal and non-problematic. It is the more problematic habits or issues we have to worry about. What are some examples?

This is not an exhaustive list but hits the main points and theme. The theme of -- if you are doing or feeling this, it is time to put down the phone and step away from tech.

- Feeling aches and pains related to phone usage. Like pain in the neck or aches in the hand you typically use your phone in.
- Feeling alone while surrounded by family or friends.
- Feeling addicted and hyper attached to your phone.
- Texting and wanting an instant response.
- Texting while driving because of boredom.
- Feeling like you cannot live without your phone.

Our phones, like other technology, are tools to make our lives easier. They should not be able to control and incite bad habits in our lives. Please, try stay in control of your relationship with health and technology.

COMING TO AN END

STAYING CURIOUS

I want to switch the tone a bit as we come to an end. The next couple of pages will be a bit free form. These handful of topics are all related to information from previous chapters, but I could not find a perfect place for them. But they are still worth discussing and adding to your cyber knowledge.

DEVICES AREN'T SPYING

Has this ever happened to you: you are driving in a car, you mention something random, then later it pops up in your social media feeds as an advertisement? Your mind might start racing, thinking, 'Siri and Alexa are spying on me, I knew it'. And I am here to tell you that nobody is spying on you.

When we sign up for subscriptions, use popular internet services or go to our favorite store with our cell phone, we are giving up data on ourselves. We typically opt into freely sharing without noticing. Data, ad, and marketing companies use this data to serve us targeted ads. They use the information of our digital avatars to do it.

It can be scary at times, how spot-on these ads can be. There are many great examples of this phenomenon, but none really beat out the story of how Target, the retail company, figured out a high school girl was pregnant before her father. It is an interesting story of targeted advertising and predictive marketing.

I strongly recommend reading an article or two on the story. To find it type this in a search engine: Target girl pregnant.

ARTIFICIAL INTELLIGENCE

Artificial Intelligence or AI is technology and machines demonstrating some behaviors normally attributed to humans -- behaviors that range from learning about the environment to understanding a conversation. The end goal for AI is creating realistic human-based pieces of technology. Something that can think, feel, and love, while one day probably becoming our supreme overlord and ruler. Joking a bit there. AI is frequently misunderstood.

There are a few different ways to conceptualize AI, but we will focus on two concepts: weak and strong. They best encompass the ideals of AI and its actual functionality.

WEAK AI

Artificial intelligence seeks to emulate the human mind. If AI, as a whole, wants to emulate the mind, then weak AI is the compartmentalization of that goal. Weak AI is an artificial intelligence that only performs limited or a focused task of the mind. In simpler terms, technology with weak AI is not conscious, but it can organize, categorize, and make accurate predictions with the data it is given. Things like Siri, Alexa and Spotify's AI music recommendations are all forms of weak AI. What is strong artificial intelligence?

STRONG AI

Strong AI is near or equal to human levels of intelligence. A machine with strong AI controlling its decisions can learn, plan, and communicate. Strong AI allows a computer to apply general collected knowledge to any task or issue presented to it. This is drastically different from weak AI which only follows a limited set rule for the data it works on. If weak AI is a simulation of human cognition, then strong AI is a true emulation.

Strong artificial intelligence is the human mind staring at a machine and not being capable of telling the difference. It is a scary concept because any technology with strong AI effectively has sentience. It can make decisions based on its own interest and not the interest of its creator.

It is just a concept for now. Strong artificial intelligence does not currently exist. But one day it will. We as humans must be prepared for this eventuality. Not being the only conscious being on Earth will be terrifying at first. But hopefully we developed and cultured our mindsets for growth and change.

RACIAL BIAS

Since the concept of race was invented. There has been racial bias and the tech world is not excluded from the reach of it.

Racial bias existed when cameras and film were being developed, i.e.: a Shirley card and other decisions and technology. A Shirley card, for anyone curious, was a film lab technicians would use to calibrate a film's colors against. What makes this process racial bias? Shirley cards contained a white woman with brown hair. Wondering why I brought up the camera?

A few years ago, facial recognition technology started to be used. Facial recognition is an amazing innovation and close to the sci-fi turned real. Opening doors, phones and starting our cars with our face is all now possible. Well if you are fairer skin. But if you are darker skinned, there might be some issues. Facial recognition has a problem with identifying darker skinned folk at the same rate as fairer skinned. And this all stems from racial bias in the cameras themselves. But it does not end there.

THIS JOB SHOULD EXIST

Cyber security awareness is not something a lot of companies put thought into. Why would they? Proving a return on investment (ROI) for it is tough.

Educating employees on the cyber risks a company faces and how to defend against them is hard to validate a need for. If the training is truly effective, then the results will be barely noticeable. When employees do everything right, it will look as though they are doing nothing at all.

While on the other hand buying the latest security devices boasting amazing results has tested and traditional ways to prove effectiveness. For example, a newly purchased security device claims to block 50 unwanted network scans per hour. It blocks 49. Saving the company money with automated defense and providing a noticeable ROI. But what about the future?

At some point, cyber security was thought to be an unnecessary investment in a company's resources. Even though at some point in our history of technology, we feared a floppy disk could destroy the world. Which seems like a laughable thing in our modern day. But for some reason, cyber security was not seen as a necessity until the early 2000s.

Educating users and employees on how to properly use technology, avoiding phishing emails, other threats, and how to mitigate them will become a new need for companies. Nothing proves this more than the increased cyber attacks brought on by COVID-19 and the corresponding lockdown. The result from millions of employees working from home on unsecured networks.

Companies should give employees and users the proper tools to fight on the frontlines. Most cyber attacks seek the easiest way in and oftentimes that is the uninformed and uneducated end user. This brings the need for a new cyber role called Cyber Security Awareness Engineer.

THE END IS THE BEGINNING

Topics related to cyber and technology seem infinite and in a way they are. Computer technology affects every part of our lives from the techniques used on the food we grow to machines used for exploring the cosmos. I would love to address them all, but this book was meant to be short and sweet. Something to stir the flames of curiosity in your human spirit and give a bit of knowledge along the way. The end of this book is, hopefully, just the beginning of a long journey in your world cyber. Until next time.

JAE TYLER – is weird, quirky, and a veteran of the U.S. Army. Passionate and serious about helping anyone learn computer technology and digital lifestyles.

www.ingramcontent.com/pod-product-compliance
Lightning Source LLC
Chambersburg PA
CBHW031349040426
42444CB00005B/235